Carbon Dioxide, Water Vapor and EARTH Climate Sensitivity: Concern or Crisis?

H. David Umphenour

DEDICATION

To knowledge at any level, wisdom at any age and Mutt.

CONTENTS

EPIGRAPHS

There are two kinds of truths: those of reasoning and those of fact. The truths of reasoning are necessary and their opposite is impossible; the truths of fact are contingent and their opposites are possible. - G. Leibniz

---

It is very advisable to examine and dissect the men of science for once, since they for their part are quite accustomed to laying bold hands on everything in the world, even the most venerable things, and taking them to pieces. - F. Nietzsche

1 INTRODUCTION

This is an attempt by the author to understand the basics of human induced climate change. As such it may, or may not, aid others in a similar position. It is written at the undergraduate level where a grasp of algebra, calculus, radiative transfer and basic thermodynamics may be helpful. Carbon dioxide and water vapor are considered in detail with additional greenhouse gases (GHGs) and other forcing entering the climate sensitivity where aerosols from coal burning appear to affect temperature. The IPCC 2C limit and geoengineering proposals for mitigation are considered. Only temperature increases of the Earth's atmosphere and surface are considered with the pre-industrial standard year being 1750. There is no attempt to deal with sea-level rise. A historiographical re-hash is not included. The reader is encouraged to review the efforts of J. Fourier, J. Tyndall, S. Arrhenius, G. Callendar, G. Plass, C. Keeling, et al., at their own discretion.

The Oxford philosopher N. Bostrom has stated machine intelligence (e.g., V. Vinge's and R. Kurzweil's "Singularity" (Tegmark)), biotechnology and nanotechnology as the three most severe threats to humans. Nation-state or rogue nuclear weapons use must also be considered. While climate change may be more likely, and its potential to influence our ecological behavior is debatable, it is less likely to exterminate our species (Marcus) although other species are in question. Over 3.5 billion years, through many changing climates, one ten-thousandth of all the species that ever existed are still around, though probably not for much longer on geologic time scales. Extinction or evolution is the rule, survival - either fit or non-fit - the brief exception.

2 RADIATIVE CLIMATE MODEL

Consider a common model for so-called greenhouse planetary climate. For Earth, approximately 20W/m² by convection and 80W/m² by evaporation/condensation in the lower atmosphere are eventually lost to space as long-wave radiation. Both losses are implicit in the author's derived radiation balance which is similar to other approaches (PSU-DMAS).

Assume an Earth of mean radius r_E with a finite vertically stratified atmospheric layer of thickness z ($<<r_E$) that is homogenous in the horizontal directions. Solar radiation of power density H impinges on a hemispheric cross-section of the Earth. A fraction a, known as the albedo, will be reflected by the Earth's atmosphere and surface. The remaining incident solar power primarily in the 0.3μm (0.0000003 meter) to 1μm (0.000001 meter) wavelength range is

$$(1 - a)H\pi r_E^2.$$

The Earth's surface at temperature T_E essentially radiates as a blackbody with a power density given by the Stefan-Boltzmann law as

$$\sigma T_E^4 \qquad \sigma = 5.67x10^{-8} \, W/m^2/degK^4.$$

Since the Earth rotates, the emitted power is over the entire Earth sphere area which is four times the hemisphere cross-section or

$$\sigma T_E^4 4\pi r_E^2.$$

The atmospheric layer at temperature T_A is not as perfect an absorber or emitter as Earth and emits only a fraction ε_A, known as the gray-body emissivity, of its black-body radiation power toward the surface as

$$\epsilon_A \sigma T_A^4 4\pi r_E^2 .$$

Also, from Kirchoff's law, the absorptivity of the layer α_A must be equal to its emissivity ε_A resulting in

$$\alpha_A \sigma T_A^4 4\pi r_E^2 = \epsilon_A \sigma T_A^4 4\pi r_E^2 .$$

Now using these solar, Earth and atmosphere components one can consider three levels in the atmosphere. At the surface of the Earth the solar incident power plus the atmospheric radiated power must equal the radiated power from the surface as

$$(1 - a)H\pi r_E^2 + \epsilon_A \sigma T_A^4 4\pi r_E^2 = \sigma T_E^4 4\pi r_E^2$$

or

$$(1 - a)\frac{H}{4} + \epsilon_A \sigma T_A^4 = \sigma T_E^4 .$$

Next, the total emitted power from the bottom and top of the atmospheric layer must be in equilibrium with that absorbed by the atmospheric layer from surface emission resulting in

$$2\epsilon_A \sigma T_A^4 = \alpha_A \sigma T_E^4 = \epsilon_A \sigma T_E^4 .$$

At the top of the atmosphere the solar incident power must equal the atmospheric radiated power plus the radiated power from the surface that passes up through the atmosphere without being absorbed as

$$(1 - a)H\pi r_E^2 = \epsilon_A \sigma T_A^4 4\pi r_E^2 + (1 - \alpha_A)\sigma T_E^4 4\pi r_E^2$$

or

$$(1 - a)\frac{H}{4} = \epsilon_A \sigma T_A^4 + (1 - \epsilon_A)\sigma T_E^4 .$$

From the three level expressions then, the atmospheric layer and Earth surface temperatures can be solved (e.g., add the surface and atmospheric top equations, solve for T_E, use the middle layer equation, solve for T_A) as

$$T_A = \left(\frac{(1-a)H}{4\sigma(2-\epsilon_A)} \right)^{1/4} = \frac{T_E}{2^{1/4}}.$$

Since the effective solar radiation is nearly constant, the Earth and atmosphere are represented by their temperatures and the atmosphere's gray-body emissivity. The problem then has been reduced to finding temperatures as a function of one atmospheric parameter.

3 ATMOSPHERIC ABSORPTION AND EMISSION

Absorption by GHGs, particularly carbon dioxide and water vapor, is now considered. Their absorption is concentration and path-length dependent. The thermal region of the atmosphere exists beyond $\approx 2.5\mu m$. Carbon dioxide has two dominant absorption bands at $4.2\mu m$ to $4.45\mu m$ and one centered at $14.98\mu m$. The first is generally in complete absorption with the latter's absorption and spectral width variable. For water vapor the $6.27\mu m$ band ranging from approximately $5.5\ \mu m$ to $7\mu m$ will generally be in complete absorption. From $7\mu m$ to near $30\mu m$ water vapor's absorption will vary continuously (Smith, Wolfe).

The gray-body emissivity of the atmosphere can be written as

$$\epsilon_A = \epsilon_0 + \frac{1}{\sigma T_A^4}\left[\epsilon\int_{\lambda_1}^{\lambda_2} P_{\lambda,T_A}d\lambda + \int_{5.5\mu m}^{7\mu m} P_{\lambda,T_A}d\lambda + \int_{7\mu m}^{10\mu m}\epsilon P_{\lambda,T_A}d\lambda + \int_{10\mu m}^{14.98\mu m}\epsilon P_{\lambda,T_A}d\lambda + \int_{14.98\mu m}^{28.57\mu m}\epsilon P_{\lambda,T_A}d\lambda\right]$$

where the Planck spectral irradiance function in terms of wavelength λ in microns and temperature in Kelvin is defined as

$$P_{\lambda,T} = \frac{37418}{\lambda^5(e^{14388/\lambda T} - 1)}\ W/cm^2/\mu m .$$

The emissivity coefficients for carbon dioxide (second term) and water vapor (last four terms) correspond to the integral wavelength ranges and are further specified in following sections. For carbon dioxide, the $14.98\mu m$

5

band spectral widths are themselves a function of carbon dioxide absorption and emission (Jursa). The water vapor continuum has been split at 7μm, 10μm and the carbon dioxide 14.98μm band center for computation.

The first term in the gray-body emissivity is an assumed constant for all absorption and emission not explicitly considered in the other five terms. This includes coincident absorption of carbon dioxide and water vapor along with some methane that generally produces complete absorption from 2.5μm to 2.9μm, any absorption from 2.9μm to 4.2μm, the full carbon dioxide absorption from 4.2μm to 4.45μm, any minor water vapor absorption from 4.45μm to 5.5μm and any water vapor absorption beyond 28.57μm (below 350cm^{-1}). All methane, nitrous oxide and ozone absorption is included in the first term (Smith, Wolfe).

4 LOWTRAN

The computer program LOWTRAN (LOW (20cm^{-1} resolution) TRANsmission) contains a no scattering absorption algorithm for all GHGs (Jursa, Smith). The first versions of LOWTRAN appeared in the 1970s.

LOWTRAN emissivity over a path for carbon dioxide and water vapor is a function of optical depth or

$$\epsilon \cong \frac{1.106}{1 + 17.434 OD^{-0.6285}} \qquad OD < 3,300$$

where the non-dimensional optical depth is the product of the wavelength dependent gas absorption coefficient β and the absorber amount w over the path

$$OD = \beta(\lambda)w \,.$$

Using LOWTRAN defined absorption coefficients (Jursa, Smith), the emissivity of carbon dioxide at 14.98μm is

$$\epsilon \cong \frac{1.106 \; 5.63 w^{0.6285}}{5.63 w^{0.6285} + 17.434} \,.$$

The emissivity of water vapor from 7μm-10μm is

$$\epsilon(\lambda) \cong \frac{1.106e^{\frac{187.82\mu m}{\lambda}}w^{0.6285}}{e^{\frac{187.82\mu m}{\lambda}}w^{0.6285} + 7.28x10^9}.$$

The emissivity of water vapor from 10μm-14.98μm is

$$\epsilon(\lambda) \cong \frac{1.106 \; 230.44w^{0.6285}}{230.44w^{0.6285} + 17.434e^{\frac{65.18\mu m}{\lambda}}}.$$

The emissivity of water vapor from 14.98μm-28.57μm (350cm⁻¹) is

$$\epsilon(\lambda) \cong \frac{1.106 \; 5.93x10^4w^{0.6285}}{5.93x10^4w^{0.6285} + 17.434e^{\frac{148.35\mu m}{\lambda}}}.$$

5 HEMISPHERICAL EMISSIVITY

At this point it must be noted that the gray-body emissivity of the atmosphere is a hemispherical value and the just defined LOWTRAN values are directionally dependent because of the absorber path w dependence on the zenith angle θ. Vertical, slant and horizontal paths all produce differing absorber amounts.

The upward hemispherical emissivity of the atmosphere at the surface is identical to the downward hemispherical emissivity at the top of the atmosphere. The hemispherical spectral emissivity can be found from the directional spectral emissivity (Wolfe) as

$$\epsilon_H(\lambda) = 2 \int_0^{\pi/2} \epsilon(\lambda) \sin\theta \cos\theta d\theta .$$

Slant path absorber amounts can be written in terms of the vertical absorber amount as

$$w \cong \frac{w_v}{\cos\theta} .$$

After a change of integration variable, the ratio of the hemispherical spectral emissivity to vertical spectral emissivity can be written using the LOWTRAN emissivity in the following form as

$$\frac{\epsilon_H(\lambda)}{\epsilon_v(\lambda)} = 2(L + M) \int_0^1 \frac{x}{L + Mx^{0.6285}} dx \qquad x = \cos\theta$$

where L and M at the waveband centers are given in the enclosed Table.

λ (μm)	L q=0.6285	M
14.98 c	$5.63w_v^q$	17.434
8.5 w	$3.95x10^9w_v^q$	$7.28x10^9$
12.49 w	$230.44w_v^q$	$3.22x10^3$
21.775 w	$5.93x10^4w_v^q$	$1.59x10^4$

6 1976 U.S. STANDARD ATMOSPHERE

To proceed further, the atmosphere must be specified to enable determination of the absorber amounts for each gas. Begin by considering the 1976 U. S. Standard Atmosphere with 330ppmV CO_2 and no water vapor with all other specific constituents (Jursa, Smith). This version varies little from other U.S. Standard Atmospheres. The molecular weight of 28.9644 g/mole remains constant up to z=87km. This atmosphere's air-Earth surface interface temperature is well known to be T_E= 288.15K (15C). The mean temperature of the atmospheric layer must then be

$$T_A = \frac{T_E}{2^{1/4}} = 242.3043K \cong -31C .$$

The air temperature from the Earth's surface up through the troposphere to the tropopause (216.65K at z=11.03km) for this atmosphere varies as

$$T = 1.055T_o(1 - (0.0225/km)z) \quad T_o = 273.15K$$

where the second term coefficient of ≈-6.5degK/km is the troposphere dry lapse-rate. The temperature above the tropopause remains a constant 216.65K up to 20km in the stratosphere.

The thickness of the atmospheric layer can be found that satisfies the average temperature over z as

$$T_A = \frac{1.055T_o\int_0^z (1 - (0.0225/km)z)dz + 216.65K(z - 11.03km)}{z} .$$

This thickness is found to be 15.37km upon solving for z after the integration.

The hydrostatic pressure for this dry atmosphere approximates an ideal gas and varies up to the tropopause as

$$P = P_o(1 - (0.0225/km)z)^{5.263} \qquad P_o = 101{,}325N/m^2 \,.$$

Above the tropopause to 15.37km in the stratosphere the pressure varies as

$$P = \frac{22{,}580}{101{,}325}P_o e^{-(0.15693/km)(z - 11.03km)} \,.$$

The dry air density is

$$\rho = 3.484x10^{-6}\frac{P}{T} \qquad g/cm^3 .$$

The column density through the atmospheric layer is

$$\rho_c = \int_0^{15.37km} \rho \, dz \cong 918 \qquad g/cm^2 .$$

Thus, a one square centimeter column 15.37km high contains ≈918 grams of dry air. This is ≈92% of the total dry atmospheric mass, or ≈8% of the dry atmosphere residing above this height to 87km is not considered.

7 H₂O VAPOR

If water vapor is added to the dry air of the prior section, the pressure will increase and the atmospheric density becomes (Rogers)

$$\rho \to \rho + 0.6\rho_{vap} = \rho + 0.6RH\rho_{sat} \cong \rho$$

where the vapor density ϱ_{vap} is a function of the saturation vapor density ϱ_{sat} temperature and relative humidity RH. However, in this context, the air density and pressure will be assumed dry even when water vapor is present. This approximation should not appreciably affect the equivalent absorber amount calculation. Naturally, the water vapor density is considered where its absorber amount is explicitly considered. The water vapor saturation density up to the tropopause (11.03km) for temperature in Kelvin is

$$\rho_{sat} = 1.472 \cdot 10^{20} T^{-6.0281} 10^{-\frac{2949.1}{T}} \quad g/cm^3$$

with RH=50% assumed throughout the troposphere (Jursa). Above the tropopause to 15.37km the vapor density will be fixed at 5ppmV in the dry air stratosphere (Jursa).

The water vapor temperatures in the troposphere and stratosphere are respectively

$$T = 1.055T_o(1 + \Delta/(1.055T_o) - (0.0225/km)z) \quad T_o = 273.15K$$

and

$$T = 216.65K + \Delta$$

where Δ is the induced uniform temperature change in the atmosphere for a corresponding change in carbon dioxide concentration relative to the 1976 U. S. Standard Atmosphere value of 330ppmV (i.e., $\Delta=0$). Only the water vapor change induced by a carbon dioxide concentration change is of importance.

8 VERTICAL ABSORBER AMOUNTS

With P and T specified, the LOWTRAN absorber amount of a gas over a path in the vertical direction can be found from (Jursa, Smith)

$$w_v = \int \rho(z)\left(\frac{P}{P_0}\left(\frac{T_0}{T}\right)^{1/2}\right)^n dz$$

with $\varrho(z)$ the absorber density, $P_o=101{,}325N/m^2$ and $T_o=273.15K$. For carbon dioxide $n=0.75$ and for water vapor $n=0.90$.

For a carbon dioxide concentration C (e.g., 0.000330) the two term absorber amount (in km) is

$$w_v = C\left[0.98\int_0^{11.03km}(1-(0.0225/km)z)^{3.572}dz + 1.09\int_{11.03km}^{15.37km}e^{-(0.1177/km)(z-11.03km)}dz\right]$$

or

$$w_v = 10.646km\ C\ .$$

For water vapor the two term absorber amount (in g/cm^2) is

15

$$w_v = 1.472 \cdot 10^{25} \cdot \frac{273.15^{0.45}}{288.15^{6.4781}} \cdot \frac{RH}{100} \cdot$$

$$\int_{0}^{11.03km} \frac{\left(1 + \dfrac{\Delta}{288.15K} - (0.0225/km)z\right)^{0.3151}}{10^{\left(\frac{10.2346}{1 + (\Delta/288.15K) - (0.0225/km)z}\right)}} \, dz \ +$$

$$0.000005 \cdot \frac{22580^2}{101325} \cdot 0.3484 \cdot \frac{273.15^{0.45}}{(216.65 + \Delta)^{1.45}} \cdot$$

$$e^{(34K/km/(216.65K + \Delta))11.03km} \int_{11.03km}^{15.37km} e^{-2(34K/km/(216.65K + \Delta))z} dz$$

or

$$w_v(\Delta = 0) = 1.073 g/cm^2.$$

9 SOLAR AND ATMOSPHERE PARAMETERS

Relevant solar radiation power density and Earth albedo values (Jursa) can be specified such that along with temperature T_A the atmospheric layer gray-body emissivity can be found as

$$H \cong \frac{1368W}{m^2} \quad a = 0.300 \quad or \quad \epsilon_A = 0.775 \,.$$

This hemispherical emissivity will be the nominal value for the 1976 U.S. Standard Atmosphere with water vapor at 50% relative humidity in the troposphere and 5ppmV in the dry air stratosphere. Approximately 22.5% of Earth surface emitted thermal radiation reaches space.

10 CO₂ CALCULATION

For carbon dioxide consider again the atmospheric layer temperature. If that temperature is differentiated with respect to the hemispherical emissivity then the result after appropriate substitution is

$$\frac{dT_A}{d\epsilon_A} = \frac{d}{d\epsilon_A}\left(\frac{H(1-a)}{4\sigma(2-\epsilon_A)}\right)^{1/4} = \frac{T_A}{4(2-\epsilon_A)} = 49.45K = \frac{\Delta T_A}{\Delta \epsilon_A}.$$

The downward radiative forcing of the 14.98μm band resulting from a change in carbon dioxide level is well known (Myhre, et al.) and generally given in terms of a reference concentration as

$$\Delta F \cong 5.35 \; ln\left(\frac{C}{C_r}\right) \quad W/m^2$$

where the 92% vertical atmosphere will have slightly less forcing than that given by the expression. The forcing is also related to the change in the second term of the gray-body emissivity of the atmosphere which can be expressed as

$$\Delta \epsilon_A \cong \frac{\Delta F}{\sigma\left(T_A + \frac{\Delta T_A}{2}\right)^4}.$$

Using the differentiated term, eliminating the emissivity change (0.0188) and using two terms of a binomial expansion finds the temperature increase for a change in carbon dioxide concentration as

$$\Delta T_A \cong \frac{T_A \Delta F}{4(2-\epsilon_A)\sigma\left(T_A + \frac{\Delta T_A}{2}\right)^4} \cong \frac{\Delta F}{4(2-\epsilon_A)\sigma T_A^3}\left(1 - \frac{2}{T_A}\Delta T_A\right)$$

or

$$\Delta T_A \cong \frac{\Delta F}{4(2-\epsilon_A)\sigma T_A^4 + 2\Delta F}T_A.$$

If the carbon dioxide concentration is allowed to double from 330ppmV (w_v=351cm) to 660ppmV (w_v=702cm) then

$$\Delta T_A \cong 0.931K \qquad \Delta F \cong 3.71W/m^2.$$

Again using the second term of the gray-body emissivity, the forcing can also be written as

$$\Delta F \cong 0.5432\int_{\lambda_3}^{\lambda_4} P_{\lambda,T_A+\Delta T_A}d\lambda - 0.4278\int_{\lambda_1}^{\lambda_2} P_{\lambda,T_A}d\lambda \qquad k = 1.498x10^5$$

$$\lambda_4 = \frac{k}{10^4 - 14.98c\delta} \quad \lambda_3 = \frac{k}{10^4 + 14.98c\delta} \quad \lambda_2 = \frac{k}{10^4 - 14.98\delta} \quad \lambda_1 = \frac{k}{10^4 + 14.98\delta}$$

where the integral coefficients are the hemispherical emissivity of carbon dioxide for the two absorber amounts. From experimental work (Jursa), doubling to 660ppmV from 330ppmV will cause the spectral absorption band half-width δ to increase by $\approx$7.5% (c=1.075 or $\approx$0.02%/ppmV). From the previous expression, the spectral absorption band half-widths can be found by numerical iteration as

$$\delta \cong 52cm^{-1} \quad 13.9\mu m - 16.3\mu m \quad 330ppmV$$

and

$$c\delta \cong 56cm^{-1} \quad 13.8\mu m - 16.4\mu m \quad 660ppmV.$$

11 H₂O VAPOR CALCULATION

The change in gray-body emissivity due to water vapor is dependent on the carbon dioxide induced temperature change (0.931K) or

$$\Delta\epsilon_A = \frac{1}{\sigma(T_A + \Delta T_A)^4}\left[\int_{5.5\mu m}^{7\mu m} P_{\lambda,T_A + \Delta T_A}\,d\lambda + \right.$$

$$1.203\int_{7\mu m}^{10\mu m}\epsilon_v(\lambda)P_{\lambda,T_A + \Delta T_A}\,d\lambda +$$

$$1.386\int_{10\mu m}^{14.98\mu m}\epsilon_v(\lambda)P_{\lambda,T_A + \Delta T_A}\,d\lambda +$$

$$\left.1.051\int_{14.98\mu m}^{28.57\mu m}\epsilon_v(\lambda)P_{\lambda,T_A + \Delta T_A}\,d\lambda\right] -$$

$$\frac{1}{\sigma T_A^{\;4}}\left[\int_{5.5\mu m}^{7\mu m} P_{\lambda,T_A}\,d\lambda + \right.$$

$$1.209\int_{7\mu m}^{10\mu m}\epsilon_v(\lambda)P_{\lambda,T_A}\,d\lambda +$$

$$1.388 \int\limits_{10\mu m}^{14.98\mu m} \epsilon_v(\lambda) P_{\lambda,T_A} d\lambda +$$

$$1.053 \int\limits_{14.98\mu m}^{28.57\mu m} \epsilon_v(\lambda) P_{\lambda,T_A} d\lambda \Bigg].$$

The coefficients preceding the integrals are the ratio of the hemispherical to vertical emissivity at the center of the wavelength integration range. All other vertical emissivity wavelength dependence in the integrals is included except for the 5.5 µm to 7µm water vapor band in saturation.

Like carbon dioxide this can be written for water vapor using the same notation as

$$\Delta\epsilon_A \cong \frac{\Delta F}{\sigma\left(T_A + \frac{\Delta T_A}{2}\right)^4} = 0.0118$$

where

$$\Delta F \cong 2.33 W/m^2 \quad w_v(\Delta = 0.931K) = 1.166 g/cm^2$$

or the air temperature increase due to the water vapor increase is

$$\Delta T_A \cong 49.45K \ \Delta\epsilon_A \cong 0.585K.$$

12 CO_2 AND H_2O VAPOR RESULTS

Doubling carbon dioxide without any water vapor then increases the air and surface temperatures 0.931C and 1.107C respectively, both of which agree well with values in the general literature (e.g., Rahmstorf).

The water vapor feed-back from the 0.931C increase will increase the air temperature an additional 0.585C. Less than 100mg=0.1ml or 0.93mm of $1cm^2$ atmospheric column water increase will produce $\approx$63% of the doubled carbon dioxide without water vapor temperature increase. The significance of water vapor as a GHG absorber is evident.

The air and Earth surface temperatures are then

$$T_A = 243.820K \ (+1.516C) \qquad T_E = 289.953K \ (+1.803C).$$

Naturally, if atmospheric water vapor varies more or less from that assumed, then the stated values will change in a similar manner.

As will be seen in the following ECS section, the Earth temperature increase calculated with all relevant forcing will be $\approx$0.5C greater than that just calculated assuming only carbon dioxide and water vapor.

13 EARTH CLIMATE SENSITIVITY (ECS)

The equilibrium climate sensitivity parameter cs which considers additional radiative forcing components (Rahmstorf) is given by

$$cs = \frac{\Delta T_E}{\Delta F - h}$$

where $h \approx 0.2W/m^2$ is the thermal inertia correction to the total forcing ΔF allowing for heat flow into the Earth's oceans. Because of its transient nature, the carbon dioxide induced water vapor feed-back in the troposphere is not considered in ECS. Condensed water vapor on aerosols is considered within cloud forcing. The following uses ECS with data taken and approximated from the 2013 NOAA Annual Greenhouse Gas Index (AGGI).

Begin with the year 1976. The Earth surface temperature has increased 0.59C from 1750-1976. The carbon dioxide forcing from 1750-1976 is $0.937W/m^2$. The carbon dioxide forcing from 1976-present is $1.223W/m^2$. The Earth surface temperature has increased 0.76C from 1976-present. One can then write (Rahmstorf)

$$cs\frac{1.223W}{m^2} = \frac{\frac{1.223W}{m^2}}{\frac{0.937W}{m^2} + \alpha - \frac{0.2W}{m^2}} 0.59C \cong 0.76C \quad \alpha = 0.213W/m^2$$

where α is all other non-carbon dioxide forcing (e.g., other GHGs, ozone, albedo, aerosols, clouds, etc.).

The carbon dioxide forcing necessary to produce a 2C (+0.65C)

23

temperature rise is

$$\cong 3.21 W/m^2 \quad 504 ppmV.$$

The temperature increase when carbon dioxide doubles from the 1750 value (277ppmV to 554ppmV) will be

$$\cfrac{\cfrac{2.763W}{m^2}}{\cfrac{0.937W}{m^2} + \alpha - \cfrac{0.2W}{m^2}} - 0.59C \cong 1.71C(+0.95C).$$

The temperature increase (2.89C) when carbon dioxide doubles from the 1976 value (330ppmV to 660ppmV) will be

$$\cfrac{\cfrac{3.71W}{m^2}}{\cfrac{0.937W}{m^2} + \alpha - \cfrac{0.2W}{m^2}} - 0.59C \cong 2.3C(+1.54C).$$

Currently at 415ppmV and increasing at $\geq$ 2ppmV per year, the 330ppmV carbon dioxide concentration will have doubled by the first third of the 22nd century.

The other GHG forcing is

$$\cong 0.631 W/m^2$$

and the remaining forcing (e.g., ozone, albedo, aerosols, clouds, etc.) is

$$\cong -0.418 W/m^2 \; (cooling).$$

Now consider 35 years later or the year 2011. The Earth surface temperature has increased 1.21C from 1750-2011. The carbon dioxide forcing from 1750-2011 is 1.818W/m². The carbon dioxide forcing from 2011-present is 0.342W/m². The Earth surface temperature has increased 0.14C from 2011-present. One can then write (Rahmstorf)

$$cs\frac{0.342W}{m^2} = \frac{\dfrac{0.342W}{m^2}}{\dfrac{1.818W}{m^2} + \alpha - \dfrac{0.2W}{m^2}} 1.21C \cong 0.14C \quad \alpha = 1.338W/m^2$$

where α is all other non-carbon dioxide forcing (e.g., other GHGs, ozone, albedo, aerosols, clouds, etc.).

The carbon dioxide forcing necessary to produce a 2C (+0.65C) temperature rise is

$$\cong 3.75W/m^2 \quad 558ppmV.$$

The temperature increase when carbon dioxide doubles from the 1750 value (277ppmV to 554ppmV) will be

$$\frac{\dfrac{1.882W}{m^2}}{\dfrac{1.818W}{m^2} + \alpha - \dfrac{0.2W}{m^2}} 1.21C \cong 0.77C(+0.63C).$$

The temperature increase (2.72C) when carbon dioxide doubles from the 2011 value (389ppmV to 778ppmV) will be

$$\frac{\dfrac{3.71W}{m^2}}{\dfrac{1.818W}{m^2} + \alpha - \dfrac{0.2W}{m^2}} 1.21C \cong 1.51C(+1.37C).$$

Currently at 415ppmV and increasing at $\geq$ 2ppmV per year, the 389ppmV carbon dioxide concentration will have doubled near the end of the 22nd century.

The other GHG forcing is

$$\cong 1.020W/m^2$$

and the remaining forcing (e.g., ozone, albedo, aerosols, clouds, etc.) is

$$\cong 0.318W/m^2 \; (warming).$$

Over 35 years one sees carbon dioxide and other GHGs increasing as expected. The Earth albedo has decreased (less cooling) and there is aerosol partial absorption warming as a result of inefficient coal burning which is a substantial part of the remaining forcing increase of

$$\cong 0.736 W/m^2 .$$

The remaining forcing probably crossed the neutral point near the end of the 20th to start of the 21th centuries. As other research also indicates, the net cloud effect is expected to be neutral or slightly positive.

In 2014 the IPCC defined a 2C increase by 2036 to 2046 as a limit to avoid dangerous climate levels. That is 0.65C above the current increase of 1.35C. This is based on the their assumption that coal burning is reduced, thereby reducing aerosol based warming, in turn requiring a lower carbon dioxide concentration to prevent a 2C rise. However, the non-IPCC calculations herein are empirically based (NOAA) and assume no coal-based aerosol or carbon dioxide reduction. Inefficient coal burning aerosol partial absorption, causing some warming, mitigates (delays) the effect of increased carbon dioxide levels, at the expense of increased pollution. The 1976 calculation yields a 2C rise prior to 2064. Similarly for the 2011 calculation, a 2C rise occurs prior to 2090. Roughly, both latter approaches approximate a temperature increase by ≈2170 of ≈1.45C above the present temperature, or ≈2.8C over the ≈420 years since 1750. One could approximate the temperature increase over the next 150 years to equal the temperature increase that has occurred over the last 270 years.

14 GEOENGINEERING (GE)

Could the ECS results be altered? How long does one have? Two areas are being considered for geoengineering the planet to mitigate climate change: 1) carbon dioxide removal and 2) effectively reducing incident solar energy.

Carbon dioxide can be removed from ambient air or power plants and stored underground. More trees could be planted or the oceans could be seeded with Fe to enrich phytoplankton. Trees and phytoplankton remove carbon dioxide and produce oxygen by photosynthesis.

Solar energy can be reduced by placing mirrors or sun-shades in space. A fine sea spray from ocean water injected into the troposphere will whiten clouds (i.e., similar to the Twomey effect) thereby increasing their solar reflectivity. Injection of sulfate aerosols by aircraft into the stratosphere also increases solar reflectivity.

So, are any of these techniques realizable? There are complex political and logistical issues involved to implement any concept. The risk of unintended, and perhaps non-correctible, consequences is high (e.g., the effect of high Fe content in the oceans). A primary concern is who or what nation-state/agency decides where, or for how long, the regulated temperature is to be set (Rees)?

15 GE EXAMPLE

Consider stratospheric sulfate aerosol injection. It is known that on regional levels climate cooling resulting from volcanic eruptions has been confirmed for this method after nature has run the experiment.

Consider again the 1976 climate sensitivity

$$\frac{0.59C}{\frac{0.937W}{m^2} + \frac{0.631W}{m^2} - \frac{0.418W}{m^2} - \frac{0.2W}{m^2}} \cong 0.62C - m^2/W .$$

As an extreme case of cooling, if stratospheric aerosols had been injected resulting in a cooling of

$$\cong -0.95W/m^2$$

then the 1976 temperature could have returned to its 1750 value.

Similarly for the 2011 climate sensitivity

$$\frac{1.21C}{\frac{1.818W}{m^2} + \frac{1.02W}{m^2} + \frac{0.318W}{m^2} - \frac{0.2W}{m^2}} \cong 0.41C - m^2/W$$

a cooling of

$$\cong -3.36W/m^2$$

could have returned the 2011 temperature to its 1750 value.

In general the radiant rate of warming decreases with increasing GHGs and other forcing and the required cooling becomes more significant the longer it is delayed. The aerosol injection and cooling magnitudes would have to be maintained and adjusted as the GHG concentrations and other forcing change with time. GHGs are long-lived, remaining in the atmosphere many years, except water vapor which only resides about 10 days.

16 EPILOGUE

In closing, reducing coal burning and carbon dioxide over twenty years and not exceeding the IPCC 2C increase limit by ≈ 2040 without significant pollution seems hopeful at best.   So, should any temperature limit be expected?   The reader is left to contemplate the climate sensitivity temperature increase when carbon dioxide changes from the current value (415ppmV) where

$$cs\Delta F_{CO2} = \frac{1.35C}{\dfrac{2.16W}{m^2} + \alpha - \dfrac{0.2W}{m^2}}\Delta F_{CO2} \cong \Delta T_E \; .$$

In what manner shall this evolve?

17 WORKS CITED

IPCC - International Panel on Climate Change - All Reports.

Jursa, A.S., Editor, Handbook of Geophysics and the Space Environment. Air Force Geophysics Laboratory, USAF/DOD, 1985.

Marcus, G., "Unknown Unknowns." In Brockman, J., Editor, What *Should* We Be Worried About? HarperCollins, NY, 2014.

Myhre, G., et al., "New estimates of radiative forcing due to well mixed greenhouse gases." Geophysical Research Letters. Vol. 25, No. 14, 2715-2718, 1998.

NOAA - National Oceanic and Atmospheric Administration - AGGI 2013.

PSU-DMAS - Pennsylvania State University - Department of Meteorology and Atmospheric Science - Energy Balance Models.

Rahmstorf, S., "Anthropogenic Climate Change: Revisiting the Facts." In Zedillo, E., Editor, Global Warming: Looking Beyond Kyoto. Brookings Institution Press, 34-53, 2008.

Rees, M., "We are in Denial about Catastrophic Risks." In Brockman, J., Editor, What *Should* We Be Worried About? HarperCollins, NY, 2014.

Rogers, R.R., A Short Course in Cloud Physics. Pergamon Press, NY, 1979.

Smith, F.G., Editor, Infrared and Electro-Optical Systems Handbook, Vol. 2, Atmospheric Propagation of Radiation. ERIM/SPIE, Ann Arbor

MI/Bellingham WA, 1993.

Tegmark, M., "Will there be a Singularity within our Lifetime?" In Brockman, J., Editor, What *Should* We Be Worried About? HarperCollins, NY, 2014.

Wolfe, W.L., Editor, Handbook of Military Infrared Technology. Naval Research Laboratory, USN/DOD, 1965.

18 INDEX

a 2, Absorber 2, 7, 9, 11, 13, 15, 19, 22, Absorptivity 2, Aerosol 1, 23-29, Agency 27, AGGI 23, Air 11-13, 17, 21-22, 27, Albedo 2, 17, 23-26, Algebra 1, Algorithm 7, Annual 23, Arrhenius, S. 1, Atmosphere 1-5, 9, 11-12, 14, 17-18, 29, Atmospheric 2-5, 11-13, 17-18, 22

Band 5-6, 18-19, 21, Binomial 18, Biotechnology 1, Black-body 2-3, Bostrom, N. 1, Burning 1, 26, 30

c 19, C 15, cs 23, Calculus 1, Callendar, G. 1, Carbon Dioxide 1, 5-7, 14-15, 18-27, 30, Century 24-26, Climate 2, 23, 25, 27-28, 30, Clouds 23-25, 27, Coal 1, 26, 30, Coefficients 5, 7, 19, 21, Coincident 6, Column 12, 22, Computer 7, Concentration 5, 14-15, 18-19, 24-26, 29, Condensation 2, Condensed 23, Consequences 27, Constituents 11, Continuum 6, Convection 2, Cross-section 2

Dangerous 25, Delays 26, Density 2, 12-13, 15,17, Derived 2, Differentiated 18, Directional 9, Directionally 9, Doubles 24-25, Doubling 19, 22, Dry 11-13, 17

Earth 1-4, 11, 17, 22-26, Ecological 1, ECS 22-23, 27, Effect 25-26, Emissivity 3-9, 17-21, Empirically 26, Equilibrium 3, 23, Eruptions 28, Evaporation 2, Evolution 1, Expansion 18, Experimental 19, Exterminate 1, Extinction 1

Fe 27, Feed-back 22-23, Forcing 1, 18-19, 23-26, 28-29, Fourier, J. 1, Function 4-7, 13

Gas 1, 7, 11-12, 15, 23, GE 27-28, Geoengineering 1, 27, Geologic 1, GHG

THE AUTHOR

H. David Umphenour, Ph.D., is a mathematical physicist and was, until his retirement in 2013, a senior civilian weapons sensor and guidance scientist for tactical and strategic missiles, munitions and small rockets with the United States Department of Defense in California. He was educated in the sciences, humanities and arts at Fort Scott Community College, Pittsburg State University, Pacific Western University and the University of Missouri (now Missouri University of Science and Technology). He is a member of the American Institute of Physics - Sigma Pi Sigma, a member of the American Geophysical Union and a senior member of the American Institute of Aeronautics and Astronautics.